CONTENTS

FOREWORD

We hope that this "how to" book will give state and local officials, as well as private industry officials, practical insights into managing the cleanup of hazardous waste sites using either state funds or money from the federal "Superfund" program. Based on our four in-depth case studies we have included a lessons-learned section with an operational guide to how governments should organize their hazardous materials programs—including, most importantly, the common pitfalls to be avoided.

The Council of State Governments has recently reorganized its research activities to emphasize the production of timely information that relates to the current and immediate concerns of state executive and legislative decision makers. We have established an institute of state government research with independent centers, such as the Center for the Environment and Natural Resources, that mirror the principal areas of concern of state government.

This publication is an example of our new orientation. It was written by J. Ward Wright with support under a grant received from the National Science Foundation.

Frank Hersman
Director of Research and State Services

INTRODUCTION

The hazardous waste problem in the United States can hardly be exaggerated. A recent report by the Congressional Budget Office (CBO) estimates approximately 266 million metric tons a year are being generated. CBO also estimates that American industries may have spent nearly $6 billion in 1983 to treat or dispose of these wastes. Most of the "disposal" is in or on land.[1] The Environmental Protection Agency (EPA) has indicated that only about 10 percent of these wastes are being disposed of in an environmentally safe manner.[2]

State governments have reacted to these threats in various ways. Going back at least to 1969, states have sought to control the manner in which industries, especially those in petrochemicals, handle their waste products. In the past 15 years, an ever-increasing number of states have passed laws of increasingly broader scope and greater specificity. This trend was accelerated by the passage by the United States Congress of the Resource Conservation and Recovery Act (RCRA) in 1976, authorizing state governments to work with EPA in monitoring and regulating ongoing hazardous waste generators. Likewise, RCRA provided for a national permitting system for the treatment and disposal of hazardous wastes. With EPA certification, Congress wanted state governments to administer this permitting process. It is likely now that virtually every state has passed some legislation in furtherance of RCRA objectives.

The Comprehensive Environmental Response, Compensation, and Liability Act (CERCLA) provided $1.6 billion in a "Superfund" to assist state governments in cleaning up existing hazardous waste sites, or "dumps." The passage of this act in December 1980 led to the eventual identification of some 800 high-priority dumps as threatening enough to qualify for federal assistance. EPA estimates that up to 2,200 sites will

eventually qualify, but the United States Office of Technology
Assessment (OTA) estimates that as many as 10,000 uncon-
trolled hazardous waste sites could qualify under EPA standards.[3] Of course, these are only a few of the sites known or
suspected in the United States. Estimates ranging from 30,000
to 200,000 are given in various publications as the number of
dumps existing, and the locations of relatively few of them are
actually known.

This report is based on a research study funded by the National Science Foundation (NSF), entitled "Risk Management
and the Hazardous Waste Problem in State Governments"
(Grant No. PRA 8202347). The grant application was submitted originally by The Council of State Governments (CSG) in
the spring of 1981 to study state governmental decision-making
in the hazardous waste field. The award was made in
December 1982, to undertake four case studies in two states.
Two of the studies were to be concerned with emergency
responses to the discovery of dangerous dumps, and two were
to focus on the manner in which hazardous waste disposal
facilities received permits.

There have been many surveys of the formal steps taken by
state governments (e.g., statutes, expenditures, organizational
arrangements) to deal with hazardous wastes, and a number of
case studies of situations such as Love Canal and Hopewell,
Virginia, which dealt with industrial pollution and its impact
on victims. However, there were no studies of how state officials located dumps, analyzed them, and responded to
threatening situations. The NSF grant provided an excellent
means of investigating these processes.

As a research methodology, a case study explores a few
organizations or events in depth. This is the only effective
method for determining in detail how decision-making processes proceed in real life. In undertaking the project, the study

team sought to ascertain what actually happened, the strengths and weaknesses of each organization studied, and the steps that might be taken to make improvements. When one extrapolates from these few studies to reach conclusions about conditions in other, similar organizations (in this case, state governments), caution must be exercised. State legislators or officials reading these case studies must ask themselves whether these conditions are applicable to their states. This is something only the reader can do. It is hoped by CSG that these examples will be found to be relevant to many other state governments.

The two states selected for the study were chosen in part because they had reputations as progressive jurisdictions which had records of contending with hazardous waste problems for many years. They were both leaders in developing legislation that conformed with RCRA requirements. One state was large and one medium-sized, and they were located in geographically different areas. Officials in both states readily agreed to cooperate in the study and did, in fact, do everything possible to enhance the quality of the studies by making top officials readily available for interviews and opening their files for documentary review. It should be added that, as each set of studies commenced, local officials and representatives of the press and industry in the two states were also quite open and giving of their time.

The states involved in the study are referred to here as the Western State and the Eastern State. Their identity is not a secret, and they are referred to by name in the research report to NSF. This booklet is an effort to set forth the materials in a more universally relevant format, and space required that the situations be simplified. On balance, it was felt that readers could assess more objectively the significance and relevance of the cases to their own situations if the actual governments remain anonymous in this booklet.

Needless to say, none of these case studies depicts an operation free of problems. This is a new, difficult field, and even the best-informed decision makers in state governments are hampered by shortcomings in funds, inadequacies in information, and the inability of scientists to provide firm answers to important questions. On the basis of additional discussions with officials in other states and a comprehensive study of the literature, members of the CSG study team are convinced that similar decision-making processes throughout the 50 states would depict at least as many problems and administrative shortcomings as the governments in the four cases discussed here. For example, an authority on state governmental reactions to hazardous waste control recently wrote that RCRA regulations posed the following problems for state legislators:

- There is a lack of financial and staff resources necessary to monitor disposal facilities and enforce the regulations.

- Within state governments, there have been administrative fragmentation and confused lines of authority.

- There has been excessive interest group influence (especially from industrial organizations) upon the policy-making institutions.

- There is a lack of political leadership, especially at the executive level.[4]

Readers of the following case studies may judge for themselves the accuracy of these general conclusions as they apply to the states studied.

The case studies in the first part of this report are relatively brief summaries of the research discussed in the larger report prepared for NSF. Each of these studies attempts to relate, more or less chronologically, the sequence of decisions that were made from the initiation of action until final resolution of the threat or granting of the permit (though all of the cases have not been finally resolved as of the date this report is being written). In the course of reviewing these events, the reader can observe the conditions and constraints within which the decisions were made.

On the basis of these four studies, conclusions are presented about the principal problems encountered by decision makers in the two states. These problems include the extended time periods involved, organizational impediments to effective programs, the impact public reactions had on the decision makers, and the paucity of information with which each decision maker had to contend. In addition, the impact of unforeseeable developments on decision-making processes is described.

In the final section, a number of recommendations are made for improving these types of administrative, policy-making, and decision-making processes. This is by no means a comprehensive program for handling hazardous waste problems. Nonetheless, it is a sweeping program that would have significant structural and procedural impacts on most state governments if adopted.

CASE STUDIES

The following four case studies provided the research base for the report to the National Science Foundation (NSF). The first two cases concern responses to hazardous waste sites (or "dumps," as the term is commonly used) which were unauthorized and appeared to pose a danger to their surrounding areas. The first involved the Western State, and the second concerned the Eastern State.

The second two cases describe permitting processes whereby attempts were undertaken to construct authorized hazardous waste disposal facilities. Again, the studies are taken from the two states.

CASE 1

The Indifferent Herbicide Company

These events occurred near one of the major cities in one of the largest American states (referred to here as the Western State). Major chemical and petroleum industries are located in this urban center. In addition, it is an area characterized by a high water table and heavy dependence on groundwater for drinking supplies.

Sometime in 1970, Malcolm Fastbuck leased approximately five acres of land along a rail siding just beyond the western borders of the city. He hired a crew of approximately 15 laborers and a foreman. The crew's job was to unload drums of various chemicals from freight cars, mix the contents into an arsenic-based herbicide, and reload the drums for shipping. The crew worked on dirt in an open area near a drainage ditch that fed into a series of bayous that crisscross the metropolitan area. This area is classified as a floodplain which is subject periodically to torrential rainstorms.

In this state, responsibility for hazardous and toxic wastes rests with the department also responsible for the adequacy and purity of the state's water resources (hereinafter referred to as the Department). Field inspectors operate out of a number of districts, each of which has several inspectors. Fastbuck's herbicide operation was not inspected regularly because it was not regarded as a generator of hazardous wastes. Theoretically, there were no "waste" byproducts of the herbicide manufacturing process in this operation. On the other hand, the company did have to report to the district office on the shipment in and out of the chemicals it used in its work.

In the early part of 1977, an inspector in the district office noticed some discrepancies in the manifests Fastbuck's company had filed with the district office. The inspector was totally unaware of the herbicide operation since it was not on his list of operations to be inspected periodically. In the light of the discrepancy in the manifests, he went to the site to discuss the matter with whomever was in charge. What he saw when he arrived alarmed him. Here was a crew of workers slopping chemicals around on a dirt surface in the immediate vicinity of a major drainage ditch. As the inspector said, "I had never seen a chemical plant operating on dirt before." The inspector informed the foreman of various laws he was violating, gave him a citation, and collected a few samples of topsoil to be sent for analysis to the state laboratory. The soil proved to be saturated with organic arsenic.

The inspector sent a memorandum to his superiors in the district office and departmental headquarters in the state capital. The Department cited the company for the discharge of contaminated materials. This led the herbicide company to write quickly back that it would construct an earthen berm around the facility to contain the contaminants on-site. The berm was completed within a few months.

The inspector had also sent a copy of his citation to the local district attorney who proceeded to prosecute the company for violating several local laws. This led to the imposition of a $350 fine in August 1977 by a county court.

In spite of these responses to his original action, the inspector kept a close watch on Fastbuck's operation. Frequent tests were made of the soil at the site and water in the drainage ditch. All test results showed heavy contamination with organic arsenic. It soon became clear that the dike was not containing the contaminated materials. Under the circumstances, there was a distinct danger to the city's water supplies. In addition, though the area of the herbicide operation was isolated, there was a school in the immediate vicinity, and several residential areas were not far away. Even more significantly, the principal path of the city's residential development was directly toward the herbicide site. By the summer of 1977, the Department felt it had a significant problem on its hands; however, it was by no means regarded as the principal one faced at that time by the Department.

Officials interviewed for this study acknowledged that their information about the extent of the problem at the outset was sketchy; however, there were no funds authorized in the state budget for undertaking a comprehensive study of the area. The Department's investigative and analytical resources were limited to occasional water and soil sampling and the services of the state's laboratories. This shortage of data was to prove to be a serious handicap in legal proceed-

ings and in determining what enforcement
measures were appropriate and sufficient.

Likewise, the Department did not have funds
for taking over and cleaning up an area of this
nature. Under state law, this was a responsibility
of the private enterprise involved. On the other
hand, little insight was required to see that
Malcolm Fastbuck did not have much invested
in this operation, so he was in a position to
walk away from the problem. These circumstances necessarily lightened the hand of enforcement. In addition, encouraging economic
development was a major state policy, so the
Department was loath to put any manufacturing
operation out of business. While this was a
small company, larger enterprises might regard
its demise as a significant signal of antibusiness
policies on the part of the state. All in all, the
state's options at this point were limited. Yet,
the Department was responsible for the quality
of the water supply, and it was not going to let
the operation continue as it had in the past.

In November 1977, discouraged by the lack of
significant progress with this problem, the executive director of the Department requested the
state's attorney general to proceed against the
company to bring its operation into line with acceptable standards for the operation of this type
of enterprise. Further periodical inspections had
found contamination in some parts of the site to
be at least 5,000 times the limits established by
the state and federal governments. The short-
term steps taken by the company had not

proven to be effective. The attorney general's office instituted action to secure an injunction giving the company 60 days to submit a plan for removing the danger, and 120 days to implement it. A temporary injunction was handed down by a state court on December 5, 1977.

In support of this action, the Department had assigned one of its engineers to develop a detailed plan for the improvement of the company's site in a manner that he felt would ameliorate the threat to the city's water supply. This plan was drawn up and became the basis for an agreement between state and company representatives to be submitted to the judge in the ongoing proceedings. On December 13, 1977, the judge incorporated this plan into the permanent injunction he handed down at that time. A plan for compliance was to be ready by January 13, 1978. The company agreed to construct dikes, pools, sheds, and other improvements which would tend to contain the contaminants on the 5-acre site.

To this point, almost one year after the operation was first discovered by the Department's inspector, virtually all of the emphasis had been placed on containment. Though early tests indicated soil contamination, little attention was paid to the implications of such saturation. To meet the Department's concern with runoff, the herbicide company proposed installing a deep-well injection system. As wastes accumulated on the site, they would be flushed down the well to depths several thousand feet below the surface. In fact, at the court appearance on the injunction, the company had proposed undertaking the improvements in phases with the injection well as the final stage. While several engineers in the Department felt this was a misuse of these types of wells, the herbicide officials were nonetheless told to prepare an appropriate permit application. The company started preparing the application in February 1978 and submitted it in September of that year.

The state, in this case, processed the application entirely separate from its enforcement activities. Under its procedures, an application for an injection-well permit is a quasi-judicial process in which lawyers for the applicants and those opposed appear before a hearing officer who subsequently reports to a board. The entire process is similar to an administrative hearing before a regulatory agency. Thus, when the herbicide company submitted an application, most of the attention of the Department shifted from enforcement and prosecution to consideration of the proposed injection well.

Given this switch in emphasis, the enforcement agencies tended to back off, pending resolution of the permitting process. This did not mean that the operation was ignored, however, and the district inspector kept a close watch on the company by continuing to take soil and water samples throughout the area of the site. In addition, during the summer of 1978, more than a year after the operation was first uncovered, attention shifted somewhat from the runoff of contaminated water to the extent of soil saturation. Even at the surface, contamination was hundreds to thousands of times greater than safe levels. It began to be apparent that eventually much of the surface soil would have to be removed from the site. However, because of the attention being given to the permitting process, no formal enforcement procedures were instituted at this time to deal with this newly appreciated problem.

A long period of inactivity ensued after the formal filing of the permit application in September 1978, as technicians in the Department reviewed each part of the application, negotiated for adjustments, and suggested changes. This process went on for a year and a half. In November 1979, a draft permit was finally issued. Under the laws, public notice would have to be given before this draft could become final, and if a hearing was requested, it would have to be held. As it happened, there were many requests for a hearing.

Meanwhile, there continued to be more seepage problems at the site. The dikes and other improvements clearly were not containing the materials since test after test showed continuing runoff of contaminants. A storm in September 1979 exacerbated the entire situation when about 14 inches of rain fell within a very short period of time. This possibility had always concerned district inspectors because such rains were not unusual and the site was on a floodplain. Even the diluted rainwater contamination tested out at about one thousand times above acceptable levels. Parts of the dikes had been washed away and all of the containment basins had overflowed. A similar disaster reoccurred in January 1980. All through 1979, reports were filed showing continuing runoffs and leaching into the drainage basin. In fact, arsenic was showing up in waterways downstream as inspections extended outward.

A new element then entered the picture. In December 1979, nearly three years after the site was recognized as a danger, the public finally learned of the threat. As has been pointed out, no injection-well permit could be issued without publishing a public notice giving interested parties 45 days to indicate their desire to be heard on the matter. This notice was placed in the city's largest daily newspaper just before Christmas in 1979. Immediately, the matter became a cause celebre. Television, radio, and newspapers quickly locked onto the story and painted dark pictures of the threat this operation posed. In a short time, much of the public appeared to be in a state of panic. The hearing was scheduled to be held in the city in March 1980. By the time of the hearing, much more public attention was being paid to the continued existence of the company than to any technical attributes of the injection well.

Even more important than citizen reaction was the awakening of city and county authorities to the extent of the danger posed by the herbicide company's activities. While the public was concerned by the proximity of the site to schools and residences, city and county officials were focusing on the threat to water supplies. Initially, the city had not paid much attention to the operation since it was outside city limits; however, the city now began a comprehensive examination of the area. This process soon revealed a significant threat to a major aquifer within two miles of the site. Since 50 percent of the city's drinking water came from aquifers, and the system was integrated, alarm soon set in. The deep-injection well was not seen as a solution by local officials. Indeed, the city's hydrologist felt if anything went wrong with the well, the results could be disastrous.

In addition to the city and county governments, water utilities in the area became alarmed. This led to a concerted, cooperative effort on the part of the city, county, and utilities to fight the permit, and this coalition participated in the hearing process until the end.

The hearing period started in March 1980, and

it extended 16 months. Many hearings were held, the last on June 30, 1981—more than four years after the site was discovered. The public hearings, held before a hearing officer, were extremely contentious. After several of these meetings in the city during August 1980, the hearing officer announced that they were being transferred to the state capital more than 150 miles away. This, of course, led to a storm of protest from the public, press, legislators, and numerous representatives of citizens' associations. The city council filed a formal resolution condemning the action. Nonetheless, the hearings were moved.

At the state capital, arguments continued month after month as attorneys for all of the parties fought over every motion and item of evidence for and against the issuance of the permit. Meanwhile, inspectors from the district office continued their examinations of the site. One inspection after another showed continuing elevation of levels of contaminants farther downstream from the manufacturing site. Meetings were held with the company's representatives, and promises were given by the company to improve the dikes and retention ponds. In March 1981, a formal promise was given to pave the entire area and vastly improve the diking system.

On November 20, 1980, a new element entered the scene. Even though the Resource Conservation and Recovery Act (RCRA) had been passed in 1976, several years passed before EPA had the regulations and administrative machinery in place to enforce the act. In November, EPA inspectors appeared at the herbicide site and found unmarked containers, contaminated surface impoundments, no written analysis plan, wastewater overflowing the dikes, a failure of the company to coordinate with local police and fire departments, no contingency plan for emergencies, no written operating record, liquid running from impoundments onto the grounds of adjacent businesses, and no protective cover on earthen dikes to minimize erosion and preserve structural integrity. In April 1981, EPA notified the herbicide company that the agency was imposing a fine of $112,000. Further, if conditions were not corrected within 30 days, fines could total $25,000 per day. However, EPA did not take further immediate action.

In May 1981, another torrential rain led to a massive overflow of contaminants. Still the manufacturing process continued. The last permit hearing was held on June 30. On September 24, 1981, the company declared bankruptcy and walked away from the scene.

By this time, the principal focus of the Department and the district inspectors had shifted from runoff to soil contamination. Borings had shown significant saturation to a depth of at least 5 feet. A staff memorandum to departmental files, dated September 8, 1981, estimated the need to remove the top 3 feet of a 5.3-acre area at a cost of more than $1,600,000. This conclusion was reached after four years' experience with this problem.

Following the demise of the herbicide company, the Department called on EPA to assist in dealing with the site. The Comprehensive Environmental Response, Compensation, and Liability Act (CERCLA) had become law on December 11, 1980, during the late stages of the herbicide company's case. Eventually, the Department made application for Superfund support, and an award was made in November 1982 for site investigation. Using these funds, contractors cleaned the surface of the site and capped it with clean soil. This was a temporary step pending study of test results. Then in April 1983, six years after the discovery of the site, consulting engineers were retained with EPA funds to undertake the first comprehensive analysis of the site. In the summer of 1984, EPA reported that more than $1,300,000 had been spent in clearing and studying the site. At that time, remedial action was estimated to cost $7,850,000.

A major problem faced by state authorities throughout this experience was obtaining a definitive scientific opinion about the threat to human life and health by the substances at the site. The Department's own laboratory tests showed early on that it was dealing with arsenic and phenol, and this information never changed significantly. On the other hand, there were conflicting opinions concerning the seriousness of such contaminants. Contractors for the herbicide company maintained that since only organic arsenic was present, it was not a serious threat to human well-being. Nonetheless, these chemicals are on EPA's list of dangerous substances, and the agency provides a standard for acceptable levels that the state agencies constantly use in their own analyses. This site surpassed acceptable standards even to depths of 300 feet.

Adding to the problem was a question of whether anything happens to arsenic in combination with phenol, especially when the arsenic is in the ground. Some "experts" stated that arsenic does not break down, while others maintained that it did. The latter maintained that contact with clay and insects could lead to compositions that could be deadly. In addition, the city and water utilities were concerned about the long-term effects of arsenic leaching into the water supply, especially into groundwater. Here again the danger posed could not be stated with certainty. As late as the summer of 1984, EPA consultants were uncertain as to the threat. Their report said that the toxicology of arsenic "is a complex subject." This was especially true of organic arsenic. They felt there was a definite threat of liver or renal damage and skin and eye irritation as the arsenic passed through the water and air.

Finally, the failure of the Department to be concerned with soil saturation in the early stages of the investigation was very significant. By the latter part of 1979, departmental engineers had begun to take borings. They found significant contamination 3 to 5 feet below the surface. Soil removal was obviously going to be a major remedial problem, and the injection well would do nothing for this aspect of the problem. By May 1983, the problem was found to be much worse. Extremely high levels were found at 35 feet. A June 1983 EPA report found contamination hundreds of times above acceptable levels at depths of 39 feet. Even water-supply wells off the property were found to have unacceptable levels of arsenic at depths of 200 to 300 feet. Remedial action in 1984 called for the removal of 56,000 cubic yards of contaminated soil on-site and 24,000 cubic yards off-site. Protective slurry walls would have to be built 40 to 50 feet deep and 3 feet thick around the entire area. The study continues.

CASE 2

The Drum Pit

The county is located on an ocean bay. One afternoon in April 1981, the county's environmental health director received a telephone call from a trailer park owner who wanted to extend his trailer park over an adjacent site he had acquired. However, he was concerned about "an old pit with junk in it." The pit emitted a strong odor. The owner proposed to fill in the pit and build on top.

The environmental health director sent two county sanitarians to the site, and they reported that the pit contained discolored water and rusted drums. They also indicated that there was a chemical smell about which residents in the area had complained in the past, and they noted that water from the pit overflowed from time to time into a small, nearby stream. With this information, the county environmental health director called the Office of Environmental Programs in the state health department.

The site in question was located in a heavily wooded area. The mobile home park contained approximately 800 residents. A church camp was nearby. Anywhere from 150 to 750 people might be there at any one time. There were also a few permanent homes across a country road from the site. Even though remote, the area is a part of the United States that has a high density. Small communities are located throughout the area.

The call to the state agency led to a visit to the site by a state inspector. He completed a form that put site data into the state system.

Water samples were taken to be sent to the state laboratories. More samples were taken in May and June 1981. When interviewed, the inspector said that the area was so overgrown that it was nearly impenetrable during the initial site visits. The inspector also noted a "very faint organic odor." There were two ponds and a few old rusting drums visible on the surface. The subsequent laboratory reports showed high concentrations of 10 to 12 organic substances. Only weak concentrations of these substances were found in the nearby stream. It was clear that these were man-made substances, and that with strong concentrations at the surface, the subsurface contamination must be much greater. Because of the nearby residences and plans for the site, the problem could not be ignored.

In June 1981, the state attorney general's office was asked to step in. This led to a meeting in August 1981 of state environmental officials, an assistant attorney general, and the owners of the site. While the owners maintained they were not responsible, in fact, their relationship to the site went back a number of years. In the 1940s, the site had been excavated for clay deposits. The excavators did not refill the pit, and for several years, garbage was dumped into it. It was also known that in the summer of 1969, drums of chemicals had been put into the pit. Off and on during this period, the same people alternately owned and sold the property. This history placed the assistant attorney general in a weak position because the earlier dumping might well

have been legal when it took place, and the
owners' responsibilities at any one time were by
no means clear. Nonetheless, the owners agreed
to hire an engineering company to study the
situation in greater depth.

In November 1981, the owners' contractor
opened the area with a backhoe and found more
drums. The season had been very dry, and the
concentrations of chemicals near the surface
were not very strong. The contractor sent
samples to a laboratory, and the results confirmed the initial findings of the state inspectors.

During the winter of 1982, one of the state inspectors undertook an investigation by talking
to residents about what happened in 1969. His
inquiries identified a nearby chemical company
as the dumper and led to interviews with several
truck drivers. The drivers said that about 150
drums had been placed in the site over a period
of a few days in 1969. The records at the county
health department did not indicate permission
had been sought for this disposition of the
drums. With this information, in the spring of
1982, state officials agreed that something had to
be done. EPA was contacted to see if Superfund
money was available.

The Comprehensive Environmental Response,
Compensation, and Liability Act (CERCLA) had
been passed in December 1980. By the spring of
1982, after a year of confusion in Washington,
state officials felt there was a chance some of the
Superfund appropriation might finally be
available. EPA's regional office was contacted in

April 1982, and one of its investigators soon joined state inspectors at the site. At this time, the state officials were able to tell the EPA representative that at least 125 drums were buried at the site and that some of the identified substances were carcinogenic. With hot weather coming on, everyone could smell the chemicals. The EPA representative saw this as a possible Superfund project. From this point on, matters were negotiated between EPA and the state department of environmental affairs.

The EPA regional office soon sent a field investigation team (FIT), which decided that this was a serious situation. The FIT visit was followed by an environmental emergency response team, including an EPA contractor, to develop the information required for the EPA "10-Point Fund Authorization Report." With this documentation, $50,000 in Superfund monies could be secured to undertake more extensive analytical work. With the subsequent approval of this application, the EPA regional office was ready to move.

It is important to note that as of June 1982, the urgency of the problem was not universally recognized. A letter from the state environmental department to the county health officer dated June 25, 1982, downplayed any serious threat of the site to nearby residents. As late as August 1982, the Center for Disease Control (CDC) in Atlanta reported to the EPA regional office that dangers were minimal. Nonetheless, CDC did feel that the drums should be excavated "at the earliest possible date."

The EPA representative, who became the on-site coordinator of the cleanup effort, did not feel that this was the classic emergency response situation. However, it was his practice to start these projects with a bang. He held a press conference on the site in mid-June 1982. With the press and nearby television stations in attendance, he explained what was known about the situation and what the cleanup process would do.

The coordinator also called a meeting of all concerned governmental agencies at a motel in a nearby town. At that time, he reached a thorough understanding about the roles everyone would play in the forthcoming project. The EPA coordinator stressed that he always sought to get these understandings with agency representatives at sites away from their normal work places. In this manner, he could achieve understandings and agreements quickly without having everything sift through the bureaucratic processes of the agencies and governments involved. With these mid-June meetings, all parties were ready to proceed with the EPA-funded phase of the project, and the public was now aware that a problem existed at the site.

This is not to say that everything had gone smoothly. The county health director would subsequently say that she was caught completely by surprise by the mid-June events. Prior to that time, she had been informed several times by the state environmental department that the con-

taminants were not very dangerous. When the state police, the media, state executives, and EPA officials moved in on the site for the news conference, she was "stunned." In fact, the county health department had planned a widely advertised evacuation exercise for that same day in which the public would practice for an emergency such as Three Mile Island. Everyone in the county health department concerned with environmental matters was involved in the exercise. The conflict of events was embarrassing to all local officials concerned.

This failure to keep the local health and environmental authorities informed proved even more embarrassing in coming weeks when the local newspaper editorially criticized the county health department for "13 months of delay." As the editorial said, "[T]he county health department did nothing to protect us." In a subsequent letter to the editor, the county health director explained that she was required by state law to turn such a matter over to state environmental agencies, and this she had done promptly.

Interestingly, even after the highly publicized kickoff of the project in June, the county health department continued to receive communications from the state environmental office and the CDC saying the situation was not serious. The CDC representative, in August 1982, reported to county health officials that there were only one or two carcinogens, and the other chemicals were considered to be more in the realm of minor irritants.

In practice, the EPA coordinator was primarily concerned with local governments in terms of civil defense assistance in case a need arose to evacuate the area and call in help from local emergency services. Thus, the coordinator worked closely with the county civil defense director in establishing and operating a "hot line" to deal with public inquiries and to develop a "temporary relocation plan." The performance of this county civil defense official subsequently received high praise from the EPA coordinator.

Following the mid-June meeting, a hurricane fence was erected around the site and heavy equipment was moved in to begin the test borings and magnetic scannings required to determine the extent of the contaminated area and

the location of drums. In addition, the first
public meeting was held in an attempt to allay
fears and concerns. At this meeting, more than
100 people appeared. As in all the other public
meetings which followed, people were worried
about their children, dogs, cows, and similar
matters. They were afraid that poison would be
passed by the cows through their milk and by
the chickens through their eggs. In addition, one
of the biggest complaints (a matter which would
be raised again and again at every subsequent
meeting) was about why it had taken over a
year from the date of the site discovery before
the project got under way. This meeting was attended by local, state, and federal officials. At
subsequent meetings, CDC technicians also often
attended. By the conclusion of the project, in
December 1982, four public meetings had been
held. By and large, the same questions were
asked and answered at each. Predictably, the
number of people in attendance generally de-
clined from one meeting to the next, with only
about 20 being present at the last.

The 10-Point Authorization Report had led to
an EPA grant of $50,000 to make a study of the
area. This study was followed by a Superfund
emergency authorization. The terms of an emergency authorization limit activities to six months
and $1 million. Thus, the project had to be com-
pleted by mid-December 1982.

During June and July, a magnetometer was
used to find the underground drums, whose
location correlated roughly with the fence line

around the approximately 1.7 acres involved.
The truck drivers were also brought to the
scene, and they confirmed the extent of the
distribution of the drums. They also stated that
from 100 to 120 drums had been dumped. Thus,
at the conclusion of the fact-finding phase of the
project, the EPA coordinator reported that
"evidence reveals that 100-120 drums (mostly
solvent wastes) were dumped years ago (1968-
1970) in an area approximately 30x60 feet and
not more than 15 feet deep." On the basis of
these conclusions, the coordinator estimated the
entire cleanup effort would cost $241,000.

By mid-August 1982, the formal application
for cleanup funds was filed with EPA. The request was for $563,300. The application set forth
"assumptions" that 100 to 150 drums were involved, the project would be completed within
30 days, and 2,000 cubic yards of dirt would
have to be removed. By late October, the application was processed and the contract with the
state signed. It authorized a ceiling of $422,000,
with work to be completed by December 16,
1982.

By November 1, 1982, work began. By November 3, 154 cubic yards of dirt and five
drums had been removed. On that date, a news
conference which was covered by newspapers
and television was held at the site.

There now began a process of discovery filled
with surprises. By November 5, 181 drums had
been removed—more than anyone had estimated
at the outset. By November 8, 483 drums were

uncovered. It was apparent that layers of drums
were involved that could not have been detected
by a magnetometer sweeping the surface. On the
basis of these findings, a report on November 8
revised original estimates to "include up to a
possible total of 500 drums." The drivers were
contacted again, and they admitted that the de-
liveries in 1969 had extended "for four or five
days" and could have resulted "in a total of 360
to 480 drums [being] dumped at this site." The
November 8 report also requested an additional
$60,000 above the contractor's ceiling for drum
removal.

By November 10, the coordinator reported,
"Heavy concentrations of drums still being
found in original suspect area. Drums are leak-
ing and in a very deteriorated condition." The
report also expressed concern about high levels
of organic chemical vapors and contaminated
wastewater. All of the work during the entire
cleanup period was hampered by heavy rains.
These conditions not only impeded the actual
work effort, they also increased significantly the
dangers inherent in the chemicals. The conditions of the drums also increased costs dra-
matically since most of the drums had to be
"overpacked"; that is, the old leaky drums had
to be placed in new, secure drums before they
could be set aside for eventual removal from the
site.

The November 11 report indicated that "a
minimum of 800 drums . . . will require over-
packing and disposal." In addition, 3,500 tons of

contaminated soil would have to be removed—
an increase of 1,500 tons over original estimates.
The report also indicated to EPA that a minimum of $150,000 would be required above the
original award.

By November 14, the extent of the problem
was fairly well delineated. The report indicated
a total of 672 drums with "sludges/liquids" and
616 uncontaminated empty drums. This total of
1,288 drums would be the final number found in
a pit where, nearly two years before, a few
drums in a pool of discolored water had been
discovered.

By the conclusion of the cleanup in December
1982, approximately $900,000 had been spent. In
addition to the removal of the 1,288 drums, 23
different chemicals were found at the site. Three
drums of PCB wastes were eventually included
in removal operations in addition to 65,000
gallons of treated water and 5,000 gallons of ig-
nitable liquids which were shipped for treatment.

The removal itself was complicated by several
adverse incidents. Rain was a constant problem.
One evening, weather conditions suddenly led to
an increase in off-site organic vapor readings to
levels in excess of safe standards. The coordinator declared a "condition orange" alert
("condition red" being the worst possible), and
the civil defense director made arrangements for
a possible evacuation. Fortunately, conditions
changed within the hour so that no evacuation
was necessary.

Finally, there were several disagreements among state environmental officials and the EPA coordinator about the extent of contaminants to be removed and the condition in which the site was to be left. These matters were settled quickly in each case, in part because the project would push the million-dollar limit in any event. The amicable manner in which the project proceeded was emphasized by a note from CDC which complimented the way federal, state, and local agencies had worked together to "insure high quality, comprehensive health protection for residents who live near the site."

By mid-December 1982, the project was completed. The drums and contaminants had been removed, the pit had been filled with clean soil, a one-foot thick clay cap was installed, grass had been planted, and a videotape had been completed which EPA and the state would use to illustrate how a Superfund project could deal successfully with these types of problems.

CASE 3*

Siting an Injection Well

Deep-well injection of hazardous wastes is one scientifically acceptable method of disposal. In the Western State, the use of these wells is common. This is a state with a large petrochemical industrial base, and engineers consider the use of these types of wells to be one of the safest, most economical approaches to dealing with hazardous wastes.

In December 1978, one of the nation's major waste disposal companies applied to the appropriate state department for a permit to drill and operate a commercial hazardous waste injection well in a county along the coast. The well would be located about 3 miles from a small city (population 9,577), which is part of an industrial belt extending south from one of the nation's major urban centers. The well area would occupy a site of approximately 1 acre near a 200-acre tract already owned by the disposal company. There was very little residential property in the area, the land being limited in usefulness to industrial and commercial purposes. The acre in question was completely undeveloped and lay on a 100-year floodplain. The general region contained about 25 percent of the state's population but constituted only 6 percent of its geographical area.

There was little about the proposal that was unusual for this state other than it would be the first well permitted by this government under

<hr>

*This presentation is a condensation of the case report written by Leslie Cole.

the Safe Drinking Water Act's Underground Injection Control Program (UIC). Also, it would be the first approved under the state's RCRA-type program.

This state has a very elaborate, quasi-judicial process for the permitting procedure. Briefly, the permitting process is as follows:

- The application is filed with the responsible state agency.

- The application is examined in detail by relevant state agencies, and a draft permit is prepared for review by a committee and submitted to a commission.

- A notice is published giving affected persons an opportunity to request a hearing, and if no request for a hearing is received, the commission makes a final decision.

- If opposition is expressed, a public hearing is held before a hearing officer.

- Following the hearings, the hearing officer makes a recommendation to the commission.

- The commission makes a decision and the permit is or is not issued.

- Dissatisfied applicants may appeal to the state courts.

In this state, the location of commercial hazardous waste facilities is largely left to the developers; however, the department has siting guidelines which should be met in the selection process. The guidelines are not mandatory. The well in question was to be a general-purpose facility capable of accepting acidic, caustic, neutral organic, and organic solvent wastes. The waste liquids would be received by tank truck shipments and unloaded into one of four separate unloading chambers before being filtered and pumped into waste storage tanks. The unloading area would be paved and curbed so as to contain spills. The well itself would be drilled to a depth of approximately 11,600 feet. Many different geological and hydrological studies and presentations demonstrated that none of the aquifers in the area would be threatened by this well.

Between December 1978 and January 1980, state technicians analyzed the application in detail. This involved several departments of the state government, primarily those responsible for water resources and public health. Within the water resources department, virtually all of the technical work was undertaken by one agency which specializes in underground injection systems. The primarily engineering and geological review resulted in numerous changes to the original permit application.

Following this study, a technical summary
was drawn up and submitted with a draft permit
to the executive committee. The committee gave
its approval on January 22, 1980. It was then
sent to the commission.

The commission, on February 27, 1980, published a public notice of the permit in local
newspapers in the area of the proposed well. In
addition, formal notices were sent to adjacent
land owners. Almost immediately, requests for a
hearing were submitted by land owners, local
officials, environmental groups, and several
members of the state legislature. Public meetings
were held in the area between May 1980 and
August 1981. More hearings were held at the
state capital between August 1981 and January
1982. A final hearing was held back in the site
area in February 1982, more than three years
after the first submission of the application.

A series of presentations was prepared by a
citizens' group which had organized especially to
fight the permit. The group raised the funds
necessary to retain an attorney and expert witnesses. The group's objections were wide-ranging; however, the following were the principal points:

- There was a potential risk to
  future oil and gas development
  in the area.

- There was a potential risk that wastes would come to the surface through faults or nearby wells.

- Hurricanes and flooding posed significant dangers.

- There were better alternatives for handling wastes that the company had not evaluated.

- There was too little information available for anyone to say with certainty that the well would be safe.

The affected county government focused its objections on the following three areas:

- The company's flood protection plan was inadequate for 100-year floods.

- There were no restrictions on receiving wastes during periods of hurricane warnings.

- The trucks bringing in waste and leaving the site would increase traffic and environmental hazards.

One major coalition formed to fight the permit came to an agreement with the company whereby the latter agreed to improve its monitoring system, include an alarm system, and otherwise install a fail-safe system which presumably would keep the company operating within prescribed standards. The hearing officers (the extended hearings went through four hearing officers) incorporated this agreement in their final recommendations to the commission that the permit be issued.

The foregoing narrative probably imposes somewhat more order and rationality on the hearing process than actually occurred over the years the application was under consideration. The objections and recommendations discussed above were quite rational. Other objections were much more subjective. Many people neither cared what technology was to be used nor were impressed by the qualifications of the company's or state's experts. They felt that no guarantee of absolute safety was possible and the best way to avoid danger was simply not to create a risk. There were admissions by people at the hearings that they never understood the scientific and technical explanations which were provided. In addition, they did not believe the hearing officers understood the information either.

It is notable that the hearings faced a wall of suspicion and distrust even though the state's own best technical people had worked with the company's technicians for more than a year before forwarding a recommendation to the commission that the permit be issued. Some mentioned Love Canal and the herbicide com-

pany (described in Case 1, above) as examples that state governments did not know what they were doing.

In addition, there was a profound feeling on the part of some that there would be little supervision of the operation once the permit was granted, and they felt the company would operate as it chose. Some pointed out that the state did not provide the funds required for an adequate inspection system. In any event, many of those attending openly stated that the state was simply an advocate for the company. They did not feel the state had considered alternative approaches adequately.

One complaint deserves special consideration. The company, the state, and the parties in opposition all had "experts" at the hearings discussing the scientific and technological factors involved with the well. Much of this testimony was in conflict, and some listeners became convinced that there was no technical certainty or consensus on this subject. (As an aside, the hearings on the proposed injection well discussed in Case 1, which was also in the Western State, resulted in the same confusion on the part of those monitoring the hearings. As if to dramatize this uncertainty, some of those attending the hearings asked why, if geologists were so smart, so many dry oil and gas wells are dug.)

In spite of the opposition, the final hearing examiner recommended that the permit be issued. Among his reasons for making this recommendation were the following:

- The fact that waste generation was increasing about 10 percent annually made it imperative that disposal facilities be provided.

- On the basis of the best scientific and technological information available, the site was geologically and environmentally sound, and the land use was consistent with the surrounding area.

- Alternative, available technologies (e.g., landfilling, lagoons, ponds) were not as environmentally acceptable as the deep well.

- On the basis of the evidence presented, there did not appear any reasonable threat to underground water resources.

- The company would provide performance bonds sufficient to correct problems should the company fail to abide by regulations.

This recommendation went to the commission on November 1, 1983. This was one month short of five years after the company's original submission of its application for the permit.

Following this recommendation, the opposing parties filed a request for a rehearing. This request was permitted to die as a matter of law. The opposing parties subsequently appealed the matter to the courts.

CASE 4*

A
Landfill
on
the
Point

The Eastern State is a significant generator of hazardous wastes. Until recent years, it had only two off-site, commercially operated landfills for waste disposal, and the larger of these closed in 1981. As a consequence, the one remaining landfill was overtaxed, and most of the waste would have to be disposed of illegally or exported out of state. These circumstances forced the state to seek additional landfill capacity.

This is a progressive state in the hazardous waste field. It has phases I and II authorization from EPA to administer the RCRA program and to permit hazardous waste facilities. Permit applications are the responsibility of a special office of waste management which is in the department of health. The procedure for securing a permit is as follows:

- A preliminary report must be submitted to the waste management office indicating where the facility will be located, the purposes for which it will be used, and the method of disposal.

- A detailed analysis of the report is undertaken by technicians in the waste management office, and the application is either denied or a draft permit is prepared.

*This presentation is a condensation of the case report written by Leslie Cole.

- If the decision is favorable, a notice is published in the state's official register and in appropriate local newspapers. In addition, copies of the notice are sent to relevant local jurisdictions and anyone requesting a copy.

- Written comments are accepted for a period of 30 days following public notice, and if sufficient interest is shown, a public hearing will be held.

- After the public hearing, written comments from the public will be considered by the waste management office for 5 days.

- The waste management office prepares a written decision on the application, and applicants are notified within 60 days following the public hearing process.

This state also had a facility siting program act, passed in 1980, which, among other things, directed the state's agency for environmental services to undertake a survey of off-site needs and to identify possible facility disposal sites. The act also created a siting board, an independent agency with the authority to overrule local zoning requirements and other local restrictions if necessary in establishing suitable sites.

In accordance with the requirements of this law, the environmental department (an agency which is part of the state's health department) retained a consultant in 1980 to assess the state's overall needs for off-site hazardous waste disposal facilities. The consultant was to consider alternative types of treatment. The report was released in August 1981 and indicated, among other things, that the state needed additional landfill capacity capable of accepting up to 200,000 tons annually. This landfill would be principally for the needs of one giant industry in the state; however, approximately 175 other industrial firms also would have a need for the landfill.

At about the same time, the environmental agency contracted with another consultant to develop facility siting guidelines. This began a process which lasted more than a year and culminated in the designation by the agency of a number of areas around the state as "candidates" which potentially could be the locations of hazardous waste disposal facilities. The process of designation included a number of public meetings and nominations by local jurisdictions of prospective sites. A set of guidelines was issued in March 1981. In general, these guidelines set forth the criteria which had to be met for a prospective site to be deemed suitable for a facility. These eliminated large areas of the state from further consideration, and ranked other areas as more or less suitable for additional study. These prospective sites were then to be passed to the state siting board for certification as eligible sites. Before this final study was to begin, however, there were to be detailed, on-site hydrological studies. The environmental agency did not have the resources for these types of studies. Alternative approaches were then being sought when, in the spring of 1981, the siting board suddenly intervened and directed the environmental agency to apply for a certificate of necessity for a site forthwith.

These preemptory instructions from the siting board stemmed from the fact that all but one of the off-site facilities operating in the state had been closed, and the need for more capacity had become urgent. As a result of this order, the process of assigning priorities to sites now focused on designating one site for immediate development.

It should not be concluded that this entire study process had gone smoothly. At the outset, the public was largely indifferent as the environmental agency met with local officials in public hearings to consider generalized siting criteria. However, when the process began to narrow down to specific, named locations, the public became very interested. When the focus shifted to finding one location, the public became active in opposition.

A delegation from the environmental agency visiting a designated candidate site was met by citizens marching to protest any further consid-

eration of the site. The resulting furor led to the intervention of the governor's office in 1981 to review the process being followed by the environmental agency. After reconsideration of all of the sites which had originally seemed to be likely candidates, the environmental agency then rejected all of the candidate sites other than a piece of land the state government already owned in the heart of the state's largest industrial area. This site extended out into a bay which led to the ocean. With this decision, the environmental agency itself became an applicant for a permit to develop this site as a hazardous waste landfill.

Before discussing the manner in which this site application was processed, it is important to review several points about the survey which had been undertaken in accordance with the 1980 legislation. The state legislature wanted quick action, but it was unwilling to appropriate much money to pay for the work. The legislature felt that $300,000 would be sufficient to screen the entire state. In addition, the work was to be completed expeditiously. These funds were inadequate to undertake the detailed geological and hydrological examinations ideally required, nor were the resources adequate for educating the public. In addition, as the possibilities for sites were narrowed down, the environmental agency unfortunately used the term "candidate" site, even though these designated areas were merely considered to be possibly "approvable." The term "candidate" proved to be such a stigma that a number of homeowners complained that they could not sell their homes in these areas even after the sites were removed from the list of areas being considered. The public furor created by the overall process resulted in much of the agency's very limited financial and personnel resources being used to react to citizens' responses and to deal with higher state agencies, such as the governor's office, which wanted to know what was going on. Complete technical analyses were not undertaken as a result of this diversion of very limited resources. In the final analysis, the entire process may unnecessarily have caused these problems to the public (e.g., fear, lower property values) since there were no funds for acquiring these sites even after they were designated. Some senior state officials retrospectively saw the entire process as little more than a paper exercise.

The decision to proceed with seeking a permit for the proposed site (an area that will be referred to henceforth as the "point" since it extends out into the bay) was based on the fact that the area was already owned by the state, it had already been used as a landfill, much geological and hydrological information was already available, and the area was industrial in nature. The site was very convenient to potential users since it was in the state's largest city and near several heavily used highways. While the site was not perfect, the water in the area was already polluted, and the state made it clear that it intended to use the site for no more than 10 years. Finally, the state did not face any opposition from the city about the proposal.

With the publication of the environmental agency's decision to develop the point as a hazardous waste site, there was predictable adverse response. The first newspaper story, together with an inaccurate map, appeared in October 1981; however, state legislators and executives had started meeting with community leaders, city officials, and representatives of environmental groups much earlier in the year when the permit application was being reviewed.

There were two principal citizens' groups concerned. The first was formed by members of a small community of 22 dwellings and a church located about 2,000 feet from the site. This community was occupied entirely by blacks and had been in existence long before the area became industrialized. As industrialization had proceeded over the years, the community had complained to public authorities of many environmental problems, especially soil and water contamination. When the extended landfill was proposed as the site of a hazardous waste dump, the community formed an "improvement association" to represent its interests at the meetings and hearings.

A second group represented another community located approximately five miles from the site. It was generally concerned with environmental threats, and it also formed a citizens' association to act on its behalf.

These groups, together with city officials, finally came to an agreement with respect to the proposed landfill. The principal features of this agreement were:

- A surcharge would be levied on each ton of waste brought to the dump, and the revenue from this charge would be divided between the city and the

community of people who
would remain living nearby.

- A nearby park would be improved.

- The members of the black community would be relocated at state expense.

Meanwhile, the environmental agency was completing work on its permit application. Since the agency did not have personnel qualified to carry out the required technical studies or to assess the environmental and socioeconomic impact of the proposed dump, it had this work undertaken by a contractor. The study was completed in April 1982 and prescribed the types of lining, collection systems, monitoring systems, and surface water controls required to meet acceptable standards. Geological conditions were seen as favorable, especially since both groundwater and surface water in the area were already at least semipolluted. From a socioeconomic standpoint, since the area was already heavily industrialized and additional municipal landfill and commercial chemical waste treatment facilities were also planned for the area, the proposed landfill would not have a significantly adverse impact on the area. Basically, since air, water, and soil in the area were already impaired, no significant additional damage was threatened by the proposed new installation.

The environmental agency had submitted the application to its hazardous waste unit in April 1982. A draft permit was completed within a month. A public hearing was held on May 20, 1982, attended by approximately 300 people. Thirty-one in attendance spoke. Since so much had already been agreed upon between the environmental agency, the city, and the two community groups, the whole process was very informal. The state explained the process for selecting the point, various technicians explained how different aspects of the site and proposed facility had been reviewed, and a review was presented demonstrating the site's compliance with all relevant regulations.

Several speakers, especially elected officials, were in opposition. Complaints ranged from concern for trucks loaded with hazardous materials going through residential neighborhoods to a concern about the lack of fire hydrants in the area. An interesting point, however, was that while the state government was responsible for regulating these types of installations, it was now proposing also to operate one. A local association that acted as a watchdog over operations potentially threatening to the environment was especially insistent that compliance with regulations required an adversarial role between the regulator and the regulated. In spite of this objection, the state agency continued as the operator and regulator.

Environmental opponents stressed the unsuitability of the geology to this type of land use, the proximity of the proposed dump to residential areas, the nearness of an important river, the possibly adverse impact on the seafood industry, potential risks to drinking water supplies, and the lack of stability of the water table in the area.

As a result of the hearing, the permit was reexamined and several adjustments were made. These included the elimination of part of the area as a dumping site, thicker clay liners in portions of the area to be used, and detailed arrangements for adequate protection against fires.

The permit was issued in November 1982, without further public hearings. The legislature appropriated $2.8 million for the construction and initial operation of the facility.

Partially to offset the expressed concerns of citizens at the hearing about the trustworthiness of the environmental agency in operating the site, an advisory committee was appointed by the environmental agency which was representative of the interests expressed at the hearing. This committee was informally to observe operations and make suggestions from time to time to the environmental agency concerning possible improvements.

The site went into operation in 1983. It was, in fact, comprised of two different sites: one for solid hazardous wastes and the other for chrome ore wastes. In April 1984, the first of these was closed. Because of the high fees required of depositors ($100 per ton), many companies found it cheaper to ship their wastes out of state.

These high fees were due, in large part, to the promises made to the community groups. During the one year it operated, the site received only about 1,000 tons of hazardous waste. It had been anticipated that up to 200,000 tons would be received each year. The lack of use was due in part from the following:

- Waste was strictly screened as it was brought in.

- Most organics and liquids were banned.

- There was some resentment that the state was operating the site rather than a private company.

With the closure, the waste disposed of at the site was removed and sent out of state. The state reserved the option of reopening the landfill if the need arises.

PROBLEMS

This project did not include a survey of state governments to determine how they dealt with hazardous wastes. In fact, a number of such surveys have been undertaken by The Council of State Governments and other public interest groups. While surveys are very useful for demonstrating the *formal* arrangements, such as laws, regulations, and organizational arrangements made by states for dealing with these problems, they do not show how the systems work in practice.

This study was to determine the extent to which states use risk management techniques (whether or not they actually employ this term) in the process of coping with these types of problems, and only case studies could provide the insights into actual operations required to understand the techniques being used.

The strength of the case study approach is that it provides a detailed view of real life and a thorough analysis of what happened in practice. The weakness of the method is that it does not tell the researcher how other individuals and organizations have acted in similar situations. Likewise, the findings describe only the particular conditions addressed in the study. One cannot even be sure that the same organization would have responded in exactly the same way again.

With these strengths and weaknesses in mind, readers must answer for themselves the following questions:

- Are the problems derived from these four cases relevant to those confronting my jurisdiction?

- Are my organization, staff, and policies similar enough to those in the cases that I can draw similar conclusions?

In other words, state officials will have to decide how relevant these cases are to their particular circumstances and how much credence should be given to the conclusions drawn from these cases.

The study team members also talked with officials from other jurisdictions and thoroughly analyzed the relevant literature. This has led them to believe that these cases are reasonably characteristic of the difficulties faced by state and local governments with significant hazardous waste problems.

It also must be emphasized that the two states chosen for the study were selected because of geographical, industrial, and governmental characteristics that made them most useful for purposes of example and comparison. Both states have reputations as progressive jurisdictions with very professional staffs, and nothing found by the study team detracted from these reputations. While the following problems could be taken by some as criticisms of these states, in fact, the study team felt that both states performed as well as any other states would be likely to under the technical, fiscal, and political circumstances virtually all states face in the mid-1980s. In short, these are descriptions of problems with the nation's systems for coping with hazardous wastes. No invidious conclusions should be drawn about the individual actors in these dramas.

As has already been demonstrated, two of the case studies reviewed how the state governments identified and dealt with existing dangerous hazardous waste dumps. The other two were concerned with the methods used to authorize and develop properly designed facilities for the disposal of hazardous wastes. Obviously, in many ways, these are very different problems. The following observations concerning problems identified by the study team do not always relate to both types. On the other hand, as will be seen in the final section, a single, unified organizational and staffing arrangement should solve both types of problems, and the state's risk management program should encompass both emergency responses and facility siting. In the following chapter, the relevance of each problem to each type of situation will be explicated.

1.
Hazard Identification

Many states have already identified more unauthorized, dangerous hazardous waste sites than they have the resources with which to deal; however, this does not relieve any state of the obligation to seek out and respond aggressively to existing dangerous conditions. On the basis of the case studies, it appears that all too often the discovery and identification of a hazardous waste site is almost a matter of happenchance.

In the herbicide company case, for example, the inspector went to the scene because of discrepancies in the company's paperwork. If he had not been attuned to the manner in which these types of manufacturing processes should be undertaken, he might well have settled the paperwork problem and left the scene. In the drum pit case, the owner's request to the county to look at a suspicious pond of water opened the case. Few of the early investigators took the matter very seriously, and the problem could well have been ignored. There is no way of ascertaining how many similar situations *are* ignored every day. More significantly perhaps, both hazards had existed for about a decade before their discovery.

2.
Shortage
of
Resources

An underlying problem plaguing state officials dealing with both emergency responses and site selections was the absence of adequate resources. Especially pertinent in this respect were inadequacies of money, staff, and scientific data.

The budgets in all cases were hopelessly inadequate. Funding shortages led to staff resources being far smaller than conditions required. In one state studied, each field inspector may have been responsible for inspecting hundreds of *known* hazardous waste generators each year. This leaves little time for looking for new sites.

Beyond sheer numbers, however, was the problem of analytical scientific resources. The state laboratories available for analyzing the samples gathered by inspectors were, in most cases, ill-suited to the range and complexity of problems likely to be encountered. Likewise, both states were seriously hampered in hiring the analysts required to study adequately suspicious circumstances when uncovered. Illustrative of the scope of this problem were the two emergency response cases. In the herbicide case, six years after its discovery, the full extent of the problem was still not known. In the drum pit case, only after cleanup was authorized and well under way did state officials have any idea of the scope of the threat with which they were faced. A project expected to clean up a 150-drum dump with a dozen chemicals turned out to be a project which had to dispose of 1,200 drums containing two dozen chemicals.

Even with the siting cases where the technical factors were under state control, analytical shortcomings plagued officials. These officials admitted they did not have the funds required to study the point in the bay to the extent prudence required. In the injection-well case, with private industry paying the bill, the state itself did not have the analytical resources to make the geological and hydrological studies required to confirm or refute the company's own studies. In both of these situations, the general public had reason to be suspicious of the "evidence" presented at the hearings concerning the safety of the facilities.

It should be noted, however, that there is also a serious state-of-the-art problem with respect to hazardous wastes with which everyone must contend. Case histories, scientific discussions, and special studies by the Center for Disease Control and others all point out that the hazards posed by many substances and combinations of substances in the ground simply are not known with a high degree of certainty. There has been little in the way of scientific experimentation with hazardous waste sites, even with respect to their impact on groundwater. Until these studies are undertaken and scientifically established standards are developed, everyone concerned will be forced to act more on the basis of inference than on scientific fact.

3.
Organizational Responsibility

Throughout all of these case studies, the focus of administrative responsibility was shifting and often unclear. This was especially the case in allocating responsibility for any failures to act. Case studies tend to document what happened when people did take actions. Failures to act usually are not made a matter of record. An inspector might simply walk away from a suspicious site. A supervisor might disregard an inspector's report. Agency heads might decide that many other problems are more urgent. As far as the study team could tell, with respect to these four studies, there was no established procedure for prescribing the necessity of action under given circumstances and clearly pinpointing who was responsible for initiating action in different types of circumstances.

The problem was exacerbated in situations such as the herbicide case where enforcement inspections, judicial prosecutions, and injection-well permit proceedings were all going on simultaneously. No one had responsibility for coordinating all activities and expediting matters in a decisive and satisfactory manner. Under these circumstances, coordination, shifting priorities, and orchestrated action is virtually impossible. For example, between September 1978, when the herbicide company filed its application for a well permit, and September 1981, when the company filed for bankruptcy, the enforcement section was constantly identifying increasingly dangerous levels of contamination. However, since the permit hearings were going on, enforcement did not make any aggressive moves to force compliance. In turn, the hearing officer for the permit would not allow any evidence in the proceedings concerning ongoing pollution. The net effect of this was that the company was given three extra years of life to continue to contaminate the environment and endanger the public.

4.
Organizational Adaptability

A major feature of hazardous waste problems, whether concerned with siting or emergency responses, is that several different types of scientific and technical people may be required to deal with a problem. In all four cases, there were needs at one time or the other for geologists, hydrologists, chemists, taxonomists, and engineers. Other disciplines also possibly could have made useful contributions to understanding and dealing with the threats and possible solutions.

Both state governments in the study constantly reiterated that they did not have the funds to undertake comprehensive studies of even one site, and they did not have the scientific and technical personnel on their departmental staffs to deal with the problems encountered. In general, the tendency was to get by with available personnel. In one case, virtually all analysts were engineers and in the other, health specialists.

Since these state agencies seldom have the funds required to retain consultants, they usually undertake the work themselves. In the case of permit applications from private companies for hazardous waste disposal facilities, the tendency is to use departmental personnel to check the work of the technicians retained by the applicant. When the time comes to justify the safety of the proposed cleanup or proposed site to the public, citizens are suspicious of what they hear. They fear the self-serving nature of findings by state departmental or applicant-hired personnel.

Interdisciplinary study with each study group's personnel complement tailored to the particular facility or site under scrutiny is desperately needed. In addition, this effort must be undertaken under circumstances that best assure accuracy, objectivity, and comprehensiveness. Only in this manner will findings be convincing to the concerned public.

5.
Public
Involvement

Since Love Canal, the public reacts with alarm at the mere mention of a suspected hazardous waste dump. State, local, and private industry officials are finding the public less and less tolerant of sustained exposure to any risk whatsoever. All studies of public attitudes find a desire for zero risk. When a proposal is made to site a properly designed hazardous waste disposal facility, the public response is "not in my backyard" (or NIMBY, for short).

With respect to facility siting cases, both states studied have legislative requirements for public hearings as soon as the permit application has been reviewed by departmental staff and found to conform technically with relevant laws. In the two siting case studies, the governments performed completely in accordance with the law.

In cases involving emergency response to discovered hazardous waste dumps, legislation is much less prescriptive. "Going public" is largely a matter of agency judgment, unless something like an injection-well permit becomes involved, as was the case with the herbicide company. In fact, neither state had any standing policy for taking emergency response cases to the public. Officials in both states waited until public

notification was unavoidable. In the herbicide
case, this was when the injection-well permit
was up for final approval; in the drum pit case,
it was just before the heavy equipment was to
be moved in to begin excavation. In the first
case, the public first learned of the herbicide
company's operations nearly three years after
the site was discovered, and in the second case,
it was more than a year after the discovery. In
both instances, many citizens were outraged at
not being notified as soon as the site was recognized as contaminated.

Public officials face an extraordinary dilemma
in cases such as these. As has been seen, neither
site appeared very dangerous at the outset. State
officials could be justifiably concerned that a
premature announcement—before the state officials themselves were sure of the nature and extent of the problem—could result in great psychological stress and unnecessary losses in nearby property values. Both of these phenomena
are well-recognized results of announcements of
the presence of hazardous waste dumps. For example, suppose officials in the herbicide case announced the possible dangers in 1977, almost as
soon as they discovered the site, and subsequent
study proved the contamination was very slight.
In the period between the announcement and the
analysis (a period of more than six years in this
case), many people would panic, sell their
homes or other nearby properties, and harbor
profound concerns for their own and their fam-
ily's health. A subsequent announcement of a

"false alarm" would certainly be resented and seen as irresponsible.

In these cases, state officials are, nonetheless, faced with the ethical imperative that if there is danger, the public has a right to know. At least, people can then decide for themselves whether or not to take action.

The main difficulty in both states, however, was the lack of standing policy to guide state officials in dealing with these problems. If a policy existed, and the public was made aware of it, at least much of the atmosphere of conspiratorial silence would have been dissipated.

6.
Availability of Disposal Sites

Many, if not most, states simply have not faced up to the necessity of providing the disposal facilities required by the amount of hazardous wastes being generated within their borders. This is an even greater problem under RCRA where no disposal sites are free from the necessity of permitting and inspection. The NIMBY complex has resulted in so much political fear that many states seem literally to wish the problem would simply disappear.

These difficulties were especially demonstrated in the case where the state developed the landfill at the point. The state went through an orderly process of plan development only to find that it could not justify *any* site to the extent necessary to make it politically palatable. And in the injection-well case, the extent and nature of much of the public opposition obviously had little to do with the objective criteria and technical studies undertaken by the owner and accepted by the state. Good or bad, such a site would increase risk and depress property values. This is one of the great dilemmas of the times.

Any state legislator must realize, however, that whether or not the sites are developed, the waste *will* be generated and it *will* be disposed of—legally or otherwise.

7.
Funding

Virtually everyone interviewed in the course of the case studies stressed the inadequacy of funding. These were not complaints about the absence of frills. They were observations about the gross discrepancies between legally assigned responsibilities and available resources. While legislators increasingly are passing the types of laws required for EPA-RCRA certification, they

are seldom willing to face up to the fiscal implications of these acts. Insufficient funds are provided to hire necessary inspection staffs, undertake the analytical work required, complete adequate comprehensive planning, acquire property
for hazardous waste disposal sites, or clean up
those sites which are dangerous but do not qualify for federal Superfund monies. As a result, a
number of observers of state hazardous waste
laws and policies have labeled these legislative
actions as "paper programs."[5]

In the instances of the two states included in
this study, this is too strong a term. Both states
acted to locate dangerous situations and respond
to them; however, no one in these states was
willing to say that their programs, as funded,
were sufficient to the problem.

8.
Risk
Management
Program

In the final analysis, underlying virtually all
organizational and programmatic shortcomings
is the failure of the state agencies and legislators
to develop and implement a comprehensive risk
management program. The nature and design of
such a program is the subject of the balance of
this report.

A RISK MANAGEMENT PROGRAM

The term *risk management* is commonly used in the insurance industry to refer to the likely cost of a particular risk. In recent years, this same term has been employed to describe the risks run by regulatory agencies in the federal government as they approve or disapprove certain products, especially drugs, for sale to the public. Over the years, federal risk management procedures have grown increasingly rigorous, and various mathematical techniques are used to assess risks; however, the principal elements of the discipline are very straightforward and adaptable to evaluation of most public risks for which officials are responsible.[6]

Some authorities refer to the overall field as *risk analysis*. Terminology aside, however, all agree that there are two principal components to the process:

1. *Risk assessment*: This is a scientific or technical analysis, in quantitative and qualitative terms, of the nature of the risk.

2. *Risk management*: This is a judgmental (or political or social) process whereby the results of the risk assessment are evaluated for their "acceptability" to the population potentially threatened and a policy is developed in response to this evaluation.

Virtually all authorities in this field agree on the above use of the term *risk assessment*; however, the second step is sometimes referred to as *risk evaluation* or *risk response*. For purposes of this report, *risk management* will be used as a more meaningful and comprehensive term.

While risk assessment is the highly specialized, technical aspect of studying situations potentially threatening to health or life, risk management refers to the overall programmatic effort to identify (or locate) risks, monitor them, consider and evaluate alternative ways to alleviate them, and otherwise control them. Control may include both the special process of dealing with each situation as it arises and the overall, comprehensive program of prevention (such as the RCRA program).

The preceding chapter discussed the problems experienced by state governments in the risk management process as evinced by the case studies. These provide the basis for this suggested program.

The following recommendations have the following three principal objectives:

1. To enable responsible officials to know the nature and extent of the threats of hazardous waste dumps, both those discovered and those proposed.
2. To assess problems in such a way that both state officials and the concerned public can

have confidence in the findings.
3. To provide an organizational arrangement that will assure the involvement of qualified personnel in decision-making processes and provide responsible direction to each project.

The following nine recommendations provide a framework for a risk management program.

Recommendation 1:
AN AGGRESSIVE PROGRAM OF HAZARD DISCOVERY SHOULD BE MAINTAINED.

No state government should be satisfied with a program that merely responds to the inquiries of the public or other governments about suspected sites. There should be standing teams available for systematically surveying all suspect areas for unauthorized hazardous waste dumps and responding swiftly to any perceived threats. Teams should also have the discretionary funds required to undertake emergency steps in analyzing the situations and ameliorating the dangers to the extent possible.

Recommendation 2:
TRIAGE PROTOCOLS SHOULD BE ADOPTED FOR THE GUIDANCE OF FIELD TEAMS.

One of the principal, potential weaknesses of any state hazardous waste program is the field inspection system. Every effort must be made to provide inspectors with the types of standards and inspection guidelines that will maximize consistent quality and effectiveness in terms of scope of investigation, depth of analyses, and accuracy in interpretation. This will require a carefully developed plan, published standards, and extensive training.

A definite, standing response plan should set forth priorities for dealing with suspect sites with specified characteristics. Thus, the presence of particular chemicals at designated threshold intensities would automatically place a site in a particular category for a prescribed response. This would assist operational field personnel in determining whether the public should be warned, what sampling techniques should be followed, who should be notified, whether police authority should be involved, and whether other extraordinary actions should be taken.

This system is vital to avoiding situations where the decision to act or not is left entirely to the discretion and instincts of field inspectors and their immediate supervisors.

Recommendation 3:
*A SYSTEM FOR RISK ASSESSMENT SHOULD
BE ESTABLISHED THAT IS ENTIRELY
SEPARATE FROM OTHER RISK MANAGE-
MENT PROCESSES.*

In many ways, the risk assessment process is
the most important step in dealing with threatening situations (such as the herbicide plant) or
assessing the safety of a proposed hazardous
waste disposal facility (such as the injection
well). This is the scientific phase of the work.
The results of risk assessment establish the basis
for all risk management activities that follow.

Neither state involved in the case studies had
a formal arrangement for risk assessment. All
analytical work was undertaken by state
employees or the staffs or consultants of permit
applicants. Regardless of the actual merits of
such assessments, they necessarily were suspect
to a concerned public. As a committee of the
National Research Council has said: "Even the
perception that risk management considerations
are influencing the conduct of risk assessment in
an important way will cause the assessment and
regulatory decisions based on them to lack
credibility."[7] The public must be convinced that
any assessment group is acting with complete
scientific objectivity independently of any possible bureaucratic or political pressures.

Recommendation 4:
*RISK ASSESSMENT SHOULD USE THE BEST-
QUALIFIED PERSONNEL AVAILABLE TO
THE STATE.*

Few state governments have the number and
types of scientists and technicians on their departmental and administrative staffs to deal with
the problems encountered in analyzing complex
chemical dumps. Nor can many states afford to
retain the consultants required to deal with each
suspect site or proposed new hazardous waste
disposal facility. Nonetheless, risk assessments
must be undertaken.

Each threat or proposed facility is unique,
though many of the same problems are encountered from one site or facility to the next.
This means that teams, or panels, of scientists
and other technical people must be formed to
analyze each site. On the other hand, even in
the larger states, risk assessment is by no means
a full-time job. Panels may be formed and disbanded as sites are discovered or proposed. Ac-
cordingly, it is recommended that each state
turn to its own universities, industries, and
similar sources to develop rosters of technical
people to be called upon, jury-fashion, to form
appropriate panels as each situation arises.

All arrangements for and operations of the
panels must inspire confidence on the part of the
public. One leading authority makes the following recommendations:

- Include critical, articulate laymen [on the panel], not just "well-meaning-but-fuzzy-brained 'humanists.'" These should be people with "breadth and competence."

- Place on the record the scientists' own sources of bias and potential conflicts of interest.

- Identify the components of their decisions as being either scientific facts or matters of value judgment.

- Disclose in detail the specific bases upon which their assessments and appraisals are made.

- Reveal the degree of certainty with which the various parts of the decision are known.

- Express findings in clear, jargon-free terms.[8]

To undertake its work effectively, each panel must be provided with the field investigators, laboratory facilities, and similar resources required to collect and analyze data. All data gathering and analytical processes required to reach a decision concerning the nature and extent of the contamination and possible threats posed to the public must be under the control of the relevant panel.

The types of activities in which a typical panel might be involved are illustrated by the following steps that might be undertaken in a case such as that of the herbicide company:

1. Inventory the wastes disposed of at the site.

2. Study the composition of these wastes.

3. Determine the physical and chemical properties, such as persistence of compounds and their solubilities.

4. Ascertain the biological properties, such as toxicity.

5. Determine the potential interaction of wastes and degradation products.[9]

This process should be documented in a formal, written report to be used as the basis for dealing with the site as long as it is a matter of public concern. Of course, as new data become available and seem to have a bearing on the earlier findings, the panel would be reconvened to consider them in light of its earlier findings.

Recommendation 5:
RISK MANAGEMENT POLICIES SHOULD BE BASED ON THE FINDINGS OF THE RISK ASSESSMENT PANELS.

The work of the risk assessment panels should be restricted solely to a determination of the danger of a threatening site or the apparent risk of a proposed disposal facility. Policies for corrective action are the responsibility of the designated risk management officials in the established state agencies. Risk management officials should be bound by the technical findings of the panels. If additional data from the site in question raise doubts about the original assessment, the panel should be reconvened to consider them and amend its original findings if warranted.

Under these circumstances, risk management officials will have firm criteria with which to work in considering alternative policies, such as Superfund eligibility, partial cleanup, permit issuance, or refusal. This will enable these officials, when explaining matters to the public and defending their policy decisions, to counter accusations that scientific findings have been tailored to favored, preconceived policies by saying they had nothing to do with these findings. This procedure may also increase the public's willingness to accept the technical conclusions as being the best analysis possible, given the state's resources, and as being completely free of political considerations.

A policy binding risk management officials to the findings of the risk assessment panels gives these officials firmer ground on which to stand in dealing with the public. Bureaucratic and political considerations *must* be matters clearly separate from the assessment process.

Recommendation 6:
AN OVERALL SYSTEM FOR ADMINISTERING RISK MANAGEMENT SHOULD BE DEVELOPED.

Any administrative system incorporating the principles of risk assessment and management necessarily must start with the existing organizational structure of each state government. Some authorities have urged creation of umbrella organizations which include every possible discipline (e.g., public health, environmental affairs, water protection, industrial regulation) that might be affected by these policies. This is impractical in many, if not most, instances, and it is unnecessary.

Several organizational goals should be sought under any plan developed. The most important of these are:

- Pinpointing responsibility for each emergency response situation (such as the two cases in this report) and each site permitting process.

- Separating risk assessment from risk management.

- Creating a system for coordinating all risk management activities statewide.

- Developing temporary "project" teams for seeing each emergency response or site permitting process through from beginning to end.

- Creating an interdepartmental, cooperative arrangement which will enable project directors to tap the personnel and analytical resources in each department as and when they are needed for each project.

All of these objectives can be achieved by the use of the well-recognized organizational concept of "matrix management." This concept was originally developed in the early 1960s by defense contractors seeking tighter control over each of their contracts and better accountability to the Department of Defense. It is an approach now nearly universal in businesses operating on a contract and/or project basis.

The recommended matrix is presented in Figure 1 in a highly generalized form. This structure does not require any fundamental reorganization of a state government, though it does entail the creation of a small coordinating unit.

Figure 1

Matrix
Organization Plan
For Emergency Responses
and Waste Disposal Projects
Governor
[Appointment]
Risk
Assessment
Panels
[Support]
[Information]
Office of
Hazardous
Waste
Typical
Functional
Departments
Environmental
Affairs
Health
Public Works
Economic
Development
Project A
Project B
Project C
Project D
Emergency
Response
Teams
Site
Development
Teams
KEY
Lines of authority
Personnel assignments
Information and support
Permanent agencies
Temporary units

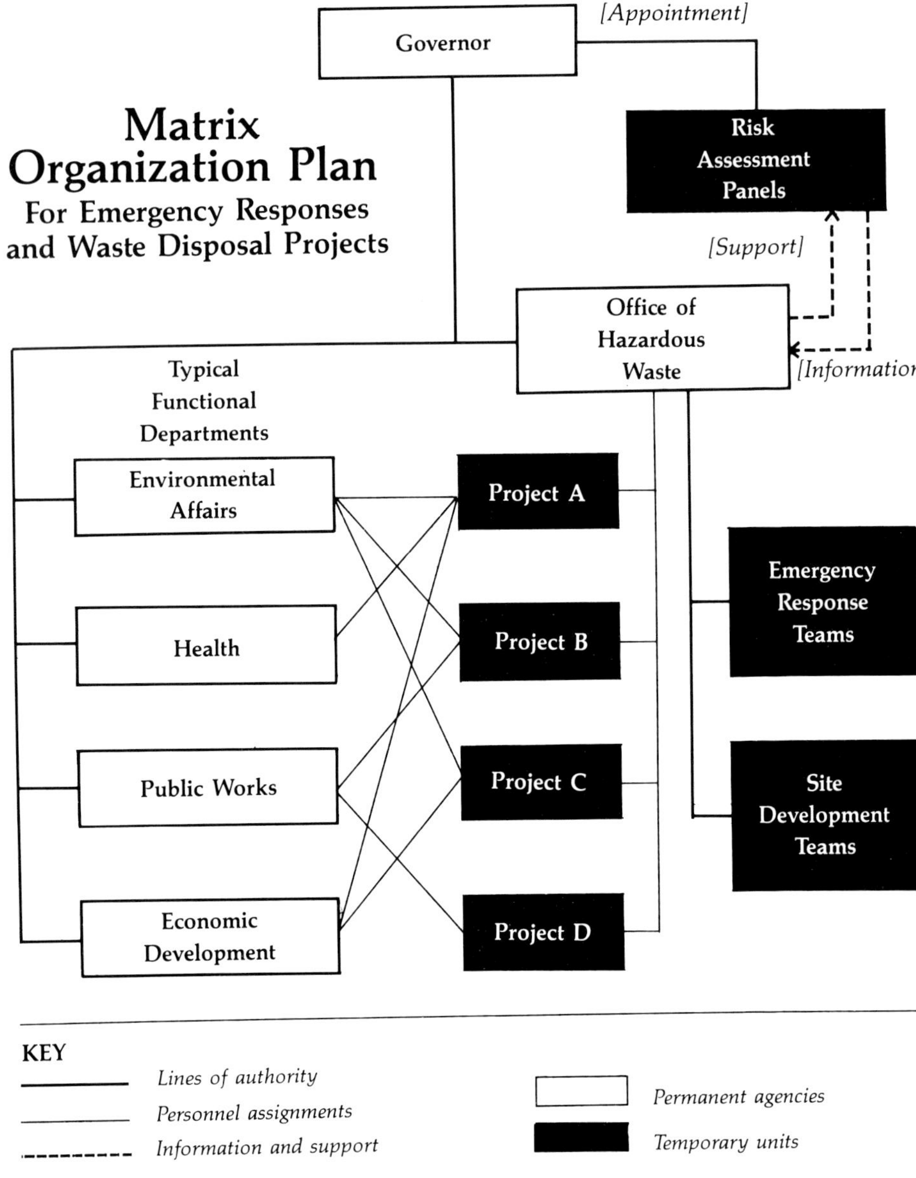

Basically, the system would work in the following manner:

- As a suspect site is discovered, or a permit to create a disposal facility is applied for, the emergency response or siting process would be treated as a temporary project.

- A project director would be selected, usually from within the state government, on the basis of technical and administrative qualifications.

- An Office of Hazardous Waste Coordination should be created to coordinate the activities of these projects. This would be a permanent agency headed by a professionally qualified director supported by as large a staff as necessary to provide oversight and support.

- As each project proceeds, a risk assessment panel would be appointed for each project when and as needed, and its activities would be coordinated and supported by the Office of Hazardous Waste Coordination.

- Each project team would be comprised of state and, if

necessary, contract employees selected on the basis of their professional or administrative qualifications and serving when and as required for project purposes. During the period of assignment, team members would report to the project director. Upon completion of their contribution to the project effort, they would return to their departments.

- State policies with respect to water protection, environmental controls, public health, and similar matters would continue to be the responsibilities of relevant state agencies, and all project members would be expected to abide by these policies to the extent that they applied to the emergency response or siting effort.

- Upon completion of each project (when the response or siting efforts are complete), the team would be disbanded and the project director would return to regular duties in a permanent department.

In any state government, the foregoing organizational and administrative arrangements should be very effective in dealing with emergency response situations. Several states have such comprehensive, formal arrangements for the review and approval of facility permits that the foregoing may not be useful. In states in which the permitting procedure is quasi-judicial, the recommended procedure would have very limited applicability. To the extent matrix management can be used, however, it provides for clear assignment of responsibilities, separation of scientific and policy responsibilities, flexibility in the use of state personnel, minimal disruption of normal state operations, and the lowest possible cost consistent with effective performance.

It should be noted that organizations using matrix management do run into many problems. The greatest of these undoubtedly is conflict between the standing departments and project managers. Arguments about the assignment of personnel, interpretations of standing policies, and overall responsibilities for various types of actions plague project directors and department heads alike. For this reason, a coordinating agency to straighten out misunderstandings and mediate disagreements is *essential*. This would be the responsibility of the proposed Office of Hazardous Waste Coordination. Of course this office would have other responsibilities, including:

- Directing the activities of emergency response teams which are charged with locating suspicious sites and taking preliminary samples.

- Creating and assisting risk assessment panels and coordinating their efforts with the efforts of the project teams they are supporting.

- Providing information and communication support to assessment panels, project teams, concerned departments, and other state officials as each participates in the overall program.

In addition, some states might see the desirability of such an office directing the state RCRA and, if it exists, the state's mini-Superfund programs.

Recommendation 7:
A COMPREHENSIVE, STATEWIDE INFORMATION SYSTEM SHOULD BE DEVELOPED.

Information is the key to the success of any effort to control hazardous wastes. RCRA requires "cradle-to-grave" information with respect to all known hazardous waste generation processes. The public demands to know as much as possible about every suspicious site or proposed hazardous waste disposal facility. However, as important as information is, as the cases in this report have shown, this is the weakest link in the chain of effective control. Much of the problem has to do with the inadequacy of scientific information in general with respect to these situations. On the other hand, even more of the problem relates to the failure on the part of most states to develop comprehensive, computer-based systems with integrated data bases which are easily accessible to all responsible state officials.

The design of such a system is beyond the scope of this study; however, it is essential. Since every state eventually will find it desirable to conform to RCRA requirements and develop a system to comply with the statute's specifications, the RCRA system may become the core of the larger state system. Included in such a system, at a minimum, are:

- RCRA tracking reports on every generator.

- Status reports on every emergency response project.

- Status reports on every proposed hazardous waste disposal facility.

- Statewide inventories and priority ratings of all known

hazardous waste sites—authorized or unauthorized.

- Content profiles of all sites that
are the subjects of emergency
response efforts.

- Administrative support reports
(e.g., budget status, personnel
assignments, target dates) for
each emergency response project.

A system of reporting requirements such as
this, based on integrated files, could provide a
data base of unlimited usefulness to everybody
from field inspectors to governors and legislators. This system is an essential feature of any
risk management program.

Recommendation 8:
*GUIDELINES FOR INFORMING AND WORK-
ING WITH THE PUBLIC SHOULD BE
ADOPTED.*

Any government official is torn between
several undesirable alternatives when trying to
deal with the public and hazardous waste problems. Especially in emergency response situations, the official faces the classic dilemma:

- "If I do not inform the residents
near the site quickly, they may
be exposed unduly to continued
exposure to contaminants; and

- "If I do inform them of possible danger and further investigation finds no significant threat, the affected public will have seen their property values fall and have suffered psychological stress unnecessarily."

Much of what has been recommended in this section has been conditioned by the need to deal effectively with the public by giving concerned citizens reason to believe that scientific studies are objective and that the state is proceeding to deal with these problems as efficiently as possible. No firm answers can be given to the questions of when the public should be informed and what they should be told; yet, whatever any given state chooses to do should be set forth in a policy statement which everybody understands and to which they must conform. This will at least remove the element of arbitrariness that now characterizes many state hazardous waste programs.

Recommendation 9:
THE STATE LEGISLATURES MUST PROVIDE THE FINANCIAL SUPPORT REQUIRED TO PUT THE FOREGOING PROGRAM INTO EFFECT.

The foregoing management program is designed in part to provide as economic an approach to this problem as possible. Much of the program may be financed under existing federal laws with EPA funds. In addition, steps may be taken to keep costs as low as possible. For example, scientists in the state may be asked to volunteer to serve on the risk assessment panels in exchange for only expenses or nominal fees; the risk management teams may be comprised of existing state employees; existing computer resources may be adequate to the needs of the information system. However, it is very unlikely that the entire program can be established without significant appropriations of state funds. Without such expenditures, states will continue in too many instances to have "paper programs."

NOTES

1. Congressional Budget Office, *Hazardous Waste Management: Recent Changes and Policy Alternatives* (Washington, D.C.: U.S. Government Printing Office, May 1985), p. xi.

2. James P. Lester, "The Process of Hazardous Waste Regulation: Severity, Complexity, and Uncertainty," in *The Politics of Hazardous Waste Management*, eds. James P. Lester and Ann O M. Bowman (Durham, N.C.: Duke University Press, 1983), p. 9.

3. CBO, *Hazardous Waste Management*, p. xii.

4. Lester, "Hazardous Waste Regulation," p. 10.

5. James P. Lester, et al., "A Comparative Perspective on State Hazardous Waste Regulation," in *The Politics of Hazardous Waste Management*, p. 212.

6. Three very informative books on risk management in the federal government are Lester B. Lave, ed., *Quantitative Risk Assessment in Regulation* (Washington, D.C.: The Brookings Institution, 1982); Committee on the Institutional Means for Assessment of Risks to Public Health, *Risk Assessment in the Federal Government: Managing the Process* (Washington, D.C.: National Academy Press, 1983); William W. Lowrance, *Of Acceptable Risk: Science and the Determination of Safety* (Los Altos, Calif.: William Kaufmann, Inc., 1976).

7. National Research Council, *Risk Assessment in the Federal Government: Managing the Process* (Washington, D.C.: National Academy Press, 1983), p. 49.

8. William W. Lowrance, *Of Acceptable Risk: Science and the Determination of Safety* (Los Altos, Calif.: William Kaufmann, Inc., 1976), pp. 110-111.

9. Richard M. Dreith, "An Industry's Guidelines for Risk Assessment," in *Risk Assessment at Hazardous Waste Sites*, eds. F. A. Long and Glenn E. Schweitzer, ACS Symposium Series No. 204 (Washington, D.C.: American Chemical Society, 1982), p. 47.

About the Author

J. Ward Wright served sixteen years with governments at the federal, state, and local levels. In addition, he worked eight years for several management consulting firms and spent several years in private law practice. In his professional career, Dr. Wright has specialized in management and financial systems. This work has included positions with Booz, Allen and Hamilton, Nassau County, New York, the National Science Foundation, the State University of New York at Albany, and The Council of State Governments.

Dr. Wright is an associate professor of management and law in the College of Business at Eastern Kentucky University. He is currently researching governmental regulation of business, especially in environmental affairs. He holds undergraduate and law degrees from the University of Chicago and a doctorate in public administration from the University of Southern California.

ORDER FORM

YES, I'm interested in *Managing Hazardous Wastes: A Programmatic Approach* (C-45)

Please send me:
_________________ copies at $15 per copy ($10.50 for state officials).

☐ Payment enclosed.

☐ Bill me.

☐ First class delivery is desired.
Postage will be added to the invoice.

NAME/TITLE _________________________________

ORGANIZATION _________________________________

ADDRESS_________________________________

_________________________ TELEPHONE: _________

CITY/STATE/ZIP _________________________________

Clip this order form and return to

The Council of State Governments
Iron Works Pike, P.O. Box 11910
Lexington, KY 40578
ATTN: Order Department, C-45

☐ Send me information on CSG mailing lists available for rent.

CSG QUESTIONNAIRE

Request: This is one of a series of CSG publications intended for your information and assistance in contending with significant state problems. It will help us to have your views on this publication. We would appreciate your effort in completing and returning this form.

MANAGING HAZARDOUS WASTES:
A Programmatic Approach
J. Ward Wright

Please mark appropriate point on each scale.

Very little — A great deal

1. To what extent has the report helped you to understand the hazardous waste problem?

   1   2   3   4   5

2. To what extent has the report helped you understand the problems of state governments in dealing with the problem?

   1   2   3   4   5

3. To what extent has the report helped you assess the capacity of your government to deal with hazardous wastes?

   1   2   3   4   5

4. Will the proposed program in the report serve the specific needs of your state government?

   1   2   3   4   5

5. What additional types of information would be of value to you in resolving your hazardous waste problems?

   ____

   ____

6. What other problems or issues concerning other areas of state government should CSG be concerned with?

   ____

   ____

Name/Title ___________________________________

Organization: _________________________________

Business Address: ______________________________

____